TROPICAL
SHIRTS & CLOTHING

Nancy N. Schiffer

Schiffer Publishing Ltd

4880 Lower Valley Road, Atglen, PA 19310

Copyright © 1998 by Schiffer Publishing, Ltd.
Library of Congress Catalog Card Number:97-80465

ISBN: 0-7643-0484-4
Printed in Hong Kong

Book design by Blair R. Loughrey

Published by Schiffer Publishing Ltd.
4880 Lower Valley Road
Atglen, PA 19310
Phone: (610) 593-1777; Fax: (610) 593-2002
E-mail: Schifferbk@aol.com

Please write for a free catalog.
This book may be purchased from the publisher.
Please include $3.95 for shipping.
Please try your bookstore first.

We are interested in hearing from authors
with book ideas on related subjects.

CONTENTS

ACKNOWLEDGMENTS

Neil Benson is the ultimate "dumpster diver," a term he tells me he did not coin but takes pleasure in as it applies to his collecting habits. He did co-found a luncheon club with similarly enthusiastic folks known as the "Dumpster Divers" (capital letters now) in Philadelphia. An accomplished professional photographer, he also makes jewelry, lamps, and art works from found objects and displays his collections in his home to enjoy every hour of every day.

Neil showed me his tropical shirts one day, and the idea hatched for a book. Thank you, Neil, for your help, generosity, and wonderfully open attitude. Your shirts are merely one expression of your far-raching influence on the people around you.

My friends in the world of shirts John King from St. Petersburg, Florida, and Cindi St. Clair continue to inspire me through their private collections, and I am grateful for their encouragement.

Label: **World Island Made in Singapore RN 18798**. White with colored map design with lettering, cotton, brown plastic buttons

TRAVEL & SOUVENIRS

The high tech world has forced many of us to discover for ourselves places of sanity where we can unwind, chill out, and behave badly. The tropics does it for some: hot weather with cooling breezes, beaches, sailing on untainted seas. Even if we have never been there, or go seldom, the thought of its being there refreshes and empowers us to get through the rest of our life with hope. We'll get there soon. And we might choose to wear some clothes that play out the relaxed attitude we seek.

In the late twentieth century, the print designs for these clothes are derived from polynesian motifs and original Hawaiian designs which became popular in the 1920s, '30s, and '40s. As casual styles in clothing spread east to the U.S. mainland with returning travelers and soldiers from the Second World War, travel industry leaders realized that clothing souvenirs could keep these special destinations in the minds of their clients a long time after the sand fell out of their shoes. General tropical designs sold well in California, Florida, and the Caribbean, and they did not have to have hula girls or the surfers from Hawaii to be appealing clothing items.

Throughout the second half of the twentieth century, tropical prints have grown in number and variety to embrace the distant places represented in this collection and far beyond. We hope you enjoy and are inspired by them, too.

Label: **King James California, Made in Hong Kong**. Map of the world in pink, blue and yellow tones, polyester, white plastic buttons

Label; **It's a Plutzer Prize winner**. Woman's dress. Black with white "Good Luck" and travel design, Rayon, zipper closure

Label: **HOM Création France Made in Italy Fabrique France F3, GB 40, D5, USA 40**. Black with blue lettering for tropical places: Puerto Rico, Sandpiper, South Florida, Santa Cruz, Surfing, Hawaii

Label: **On the Brink, Made in China RN 81029**. Dark blue background with print shirts and palm fronds design, silk, blue dyed shell buttons

SOUTH PACIFIC

Label: **Jockey Made in Taiwan R.O.C. WPL 13040.** White with blue treasure map and pirates design, cotton, white plastic buttons

Label: **Merona.** Dark blue background, lettering Melbourne, Reef, cotton, white porcelain buttons

Label: **Jantzen Made in U.S.A.** Swim trunks. Blue crackled background with fragmented geometric designs, cotton

Label: **Newport Blue.** Blue with white map of Tahiti and Papeete and floral design, cotton, blue plastic buttons

Label: **Terranova, A New Destiny, A New World, 100% Rayon, Made in Indonesia RN 80791**. Colorful exotic design with people and bananas, white plastic buttons

9

POLYNESIAN TAPA DESIGNS

Label: **Lauhala Made in Hawaii**. Lady's tea timer shirt. Green and white vertical leaf design. Cotton, coconut shell buttons.

Label: (block) **The Kahala made in Honolulu.**
Brown tapa design with sea shells. Cotton.
Coconut shell buttons on the left side for a lady.

Label: **Fashioned by Hukilau Fashions Honolulu RN 43220.** Purple and brown vertical tapa design, cotton, white plastic buttons

Label: **Ui-Maikai Made in Hawaii.** Green tapa and floral deign, cotton, molded metal buttons

11

Label: **Made in Hawaii**. Brown tapa and stylized pineapple design, cotton, molded plastic Asian writing buttons

Label: **Waikiki Sports**. Brown fragmented tapa and figures design, cotton, coconut shell buttons

12

Label: **Waikiki Holiday Made in Korea RN 26818**. Brown tapa, map and floral print, polyester, brown plastic buttons

Label: **Kamehameha made and styled in Hawaii.** Blue tapa and floral design, cotton, coconut husk buttons

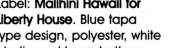

Label: **Malihini Hawaii for Liberty House.** Blue tapa type design, polyester, white plastic and brass buttons

13

Label: **Richard Douglas Honolulu Made in Hawaii.** Zippered front closure, brown and yellow border print, tapa and floral design, cotton bark cloth, two molded plastic Asian writing buttons

Label: **Sears Sportswear for pool, beach and patio, Sears, Roebuck and Co., U.S.A.** Brown and green fragmented tapa and floral design, cotton, white plastic buttons

14

Label: **Sears Sportswear for pool, beach and patio Sears, Roebuck and Co., U.S.A.** Tapa and floral design in blue and green tones, cotton, white plastic buttons

Label: **Ross Sutherland Waikiki Hawaii.** White with stylized green and blue tapa design, slit pocket, heavy cotton, white plastic buttons

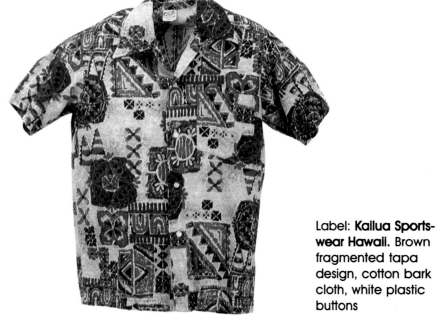

Label: **Kailua Sportswear Hawaii.** Brown fragmented tapa design, cotton bark cloth, white plastic buttons

Label: **Iolani Hawaii.** Shaded blue tapa and floral design, cotton bark cloth, fabric covered buttons

Label: **Ui-Maikai RN 26300, Made in Hawaii.** Purple and bright colored border print, cotton bark cloth, molded plastic buttons

Label: **Kai Nani Hawaii.** Shaded blue background with swirls, shoulder epaulets, waist band, and four patch pockets with buttoned flaps, cotton bark cloth, silver molded plastic Asian writing buttons

Label: **kona*kai Hawaiian Casuals Made in U.S.A. by jantzen.** Swim trunks. Shaded red and brown geometric and floral design, cotton

16

Label: **kona*kai Hawaiian Casuals Made in U.S.A. by Jantzen.** Swim trunks and shirt set. Shaded blue tapa design, Rayon, blue plastic buttons

Label: **kona*kai Hawaiian Casuals Made in U.S.A. by jantzen.** Swim trunks. Orange and brown fragmented tapa design, cotton

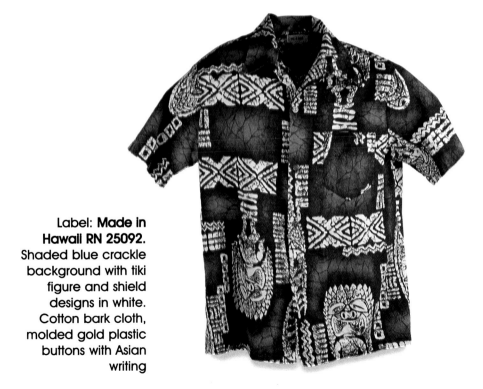

Label: **Made in Hawaii RN 25092.** Shaded blue crackle background with tiki figure and shield designs in white. Cotton bark cloth, molded gold plastic buttons with Asian writing

17

Label: **kona*kai Hawaiian Casuals Made in U.S.A. by jantzen.** Swim trunks and shirt set. Blue, green and brown tapa design, cotton

Label: **Made in Honolulu for The Aloha Shops Inc. palm beach atlantic city.** Shaded blue tapa and floral design, cotton, coconut shell buttons

Label: Royal Hawaiian. Blue background with pink tapa fragments and pink and blue hibiscus flowers, cotton, molded plastic basket weave buttons

18

Label: **Made in Honolulu for the Aloha Shops Inc. palm beach atlantic city.** Swim trunks. Blue and white tapa and floral design, cotton

Label: **The Kahala Made in Honolulu for M. McInerny.** Swim trunks. Yellow and brown tapa design, cotton

Label: **Dael's.**
Colorful Chinese
lanterns over grey
and black dragon
design, Rayon,
wooden buttons

Label: **Kai Nani
Hawaii.** Red
with Japanese
design includ-
ing cranes,
polyester, white
plastic buttons

ASIAN DESIGNS

20

Label: **Tori Richard Honolulu Made in U.S.A.** Red background and Asian writing with colorful Chinese landscape in a fan-shaped reserve and a sailboat in a rectangle, polyester, white plastic buttons

Label: **Windsor Shirt Company, Designed by the Windsor Group 100% cotton, Made in British Hong Kong RN 36543.** White background with colorful map, shell and Hotel Colombia design, cotton, white plastic buttons

Label: **Penney's 100 % Hand Washable Made in Japan.** Blue background with brown eagles and pine trees design, Rayon, white plastic buttons

21

Label: **Champion Westerns Permanent Press**. Western styling with curved shirt tail and long sleeves. Shaded brown background with tropical scene of palm trees and houses on stilts, polyester and cotton blend, Scovill snap closures

22

Label: **Boss Made in Hong Kong**. White with crests labeled Hong Kong United Kingdom, cotton, white plastic buttons

Label: (blue) **Surfriders Sportswear Manufacturers Made in Honolulu Hawaii.** Red background with multi-colored small Asian scenes and Asian writing, Rayon, red plastic buttons

HAWAIIAN MAPS

Label: **Hawaiian Holiday Sportswear Made in Hawaii.** Red labeled scenes of ancient Hawaii in horizontal bands, cotton, molded plastic Aloha Hawaii buttons

Label: **Made in Hawaii for Andrade Honolulu.** Brown Discover Hawaii design, coarse natural cotton, brown wooden buttons

24

Label: **marc daniels Made in U.S.A.** Dark blue with Hawaiian Islands map design and lettering Hawaii, with drawings of local life, cotton, blue plastic buttons

Label: **Cook Street Honolulu. Made in U.S.A. 55% Poly 45% cotton RN 20957.** Denim blue with white printed design of Hawaiian history including the ship *Resolution*, brown plastic buttons

Label: **Tropicana Hawaii**. Border print with turquoise flowers and map of Hawaii with colorful figures, cotton bark cloth, white plastic buttons

Label: **Island Fashions Made in Hawaii U.S.A.**
White with blue map and nautical design "The Principal Hawaiian Islands," 35% cotton, 65% polyester, white plastic buttons

25

Label: **Fashioned by Hukilau Fashions Honolulu RN 43226.** Shaded brown background with Hawaiian map, fish and tiki design, polyester, white plastic buttons

Label: none. Green map of Hawaii with islands labeled, flowers, boats and gold paint detail, cotton, stamped metal buttons

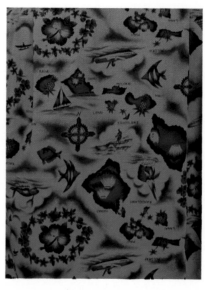

Label: **Ui-Maikai RN 26300, 100% All cotton Made in Hawaii.** Turquoise with map of Hawaii, fish and floral design, gold painted details, molded plastic basket weave buttons

Label: **Ui-Maikai RN 26300 Made in Hawaii.** Red with map of the Hawaiian islands and airplane and gold paint details, cotton, molded plastic basket weave buttons

Label: **Designed in Hawaii Calabash products wash & wear 65% polyester 35% cotton RN 37359 Made in Taiwan R.O.C.** Blue shaded background with map of Hawaii and colorful small floral design, white plastic buttons

Label: **Made in Hawaii RN 20942.** Red with map of Hawaiian islands and scattered floral and tiki designs with gold paint details and lettering, cotton, white plastic buttons

Label: **Ocean Pacific.** Swim trunks. Purple sea with labeled map of Hawaii and palm trees, cotton

27

Label: **Shoreline Hawaii.** Blue background with map of Hawaii and scattered floral and tapa designs, polyester, white plastic buttons

28

Label: **Fashioned by Hukilau Fashions Honolulu RN 43220.** Beige with brown map of Hawaii design, polyester blend, white plastic buttons

Label: **Made in Hawaii.** Green with map, anchors and Chinese boats design, cotton bark cloth, white plastic buttons

Label: **Orchid Fashions Made in Hawaii**. White with turquoise map of Hawaii and Hawaiian scenes, 65% polyester 35% cotton, white plastic buttons

Label: **Kaush Hawaii**. Turquoise with white map of Hawaii, cotton polyester blend, wooden buttons

29

Label: **New Horizon Made in Hawaii**. Child's shirt. White with blue map of Hawaii and local sights, cotton, white plastic buttons

Label: **New Horizon Made in Hawaii**. Child's shirt. White with red map of Hawaii and local sights design, cotton, white plastic buttons

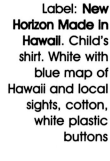

30

Label: **Helena's Made in Hawaii**. Child's shirt. White with brown map of Hawaii and landscape scenes and orange lettering "Pacific Islands," cotton, white plastic buttons

Label: **Nui Nalu Hawaii Made in U.S.A. RN 36708**. Child's shirt. White with blue map of Hawaii and local sights, cotton, white plastic buttons

Label: **Shoreline Hawaii Made in Hawaii.** White with blue map of "The Islands of Hawaii" labeled, cotton, white plastic buttons

Label: **Cook Street Honolulu.** White with blue lighthouses of Hawaii, cotton and polyester blend, brown plastic buttons

31

HAWAIIAN EMBLEMS

Label: **Malihini Hawaii Designer's Collection liberty house of Hawaii**. Orange crackle background with feather helmets, staffs and shields design, polyester, white plastic buttons

Label: **Made in Hawaii**. Red with floral and Hawaiian crest design and gold painted details, cotton, molded plastic Asian writing buttons

Label: **Tropicana Hawaii**. White crackled background and blue Hawaiian emblem crest, cotton bark cloth, white plastic buttons

Label: **Go Barefoot in Paradise Inc. Honolulu, Hawaii.** Light green crackle background with Hawaiian crest and drums design, cotton, green plastic buttons

33

34

Label: **Penneys Hawaii**. White with red horizontal print of Hawaiian crest and native warriors, cotton bark cloth, white plastic buttons

Label: **Fashioned by Hukilau Fashion Honolulu RN 43220.** Mottled brown background with tiki, drum and leaf design, cotton bark cloth, molded plastic buttons

Label: **Tropicana Made in Honolulu.** Blue crackle background with tiki and drums and gold painted details, cotton, white plastic buttons

HAWAIIAN PEOPLE

Label: **Kuu-Ipo Made in Hawaii.** Fabric design of a Hawaiian woman by Frank MacIntosh originally for a menu cover for Matson Lines cruise ships, on woman's "tea-timer" shirt, Rayon, coconut buttons

Label: (blue) **Surfriders Sportswear Manufacturers Made in Honolulu Hawaii.** "Holomu" dress of blue background and Hawaiian people design, Rayon

Label: **An Original For Watumull's by Diane Honolulu**. Dress with "Beach Boy" print fabric design in red with faces and lettering, Rayon

Label: none. Blue with border print of spear fishermen and palm trees, cotton, white plastic buttons

37

Label: **Carlton Hill RN 39012 Made in Korea**. shaded brown background with islands and spear fishermen, polyester, brown plastic buttons

Label: **Honolulu Gold**. Blue background with labeled "I love the girls of Hawaii" design, cotton, wooden buttons

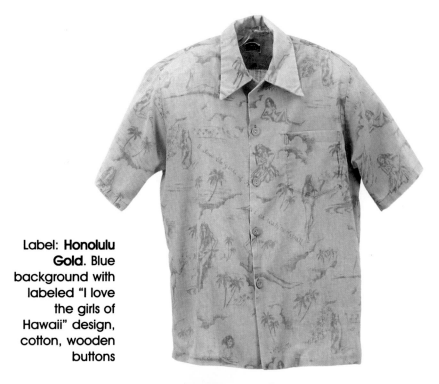

Label: **Sea Island Sportswear.** Yellow background with colorful fish and surfboard rider design, heavy cotton, white shell buttons

Label: **Michael Gerald Ltd. Hand screened Tailored in China, Made in China RN 63925**. Light blue background with pastel grass huts and hula dancers scenes, Rayon, white plastic buttons

Label: none. Orange crackle background with vertical stripes of hula dancers, floral and geometric designs, Rayon

Label: **Sea Island Sportswear**. Blue background with fish and surfboard rider design, heavy cotton, white shell buttons

39

Label: **Sundek**. Red with colorful reserves of hula dancer and ukulele player, Rayon, brown plastic buttons

Label: **Created in Hawaii for Diamond Head Sportswear Traditional Market Place Waikiki.** Green border print with black surfboard riders, cotton bark cloth, molded plastic Asian writing buttons.

Label: **Made in Korea.** Beach scene and net fishermen, polyester, brown plastic buttons

40

Label: **Mukai Fashions Made in Hawaii.** Zipper front closure, blue and white design with abstract figure in brown, cotton bark cloth, one silver plastic button

HAWAIIAN FLORALS

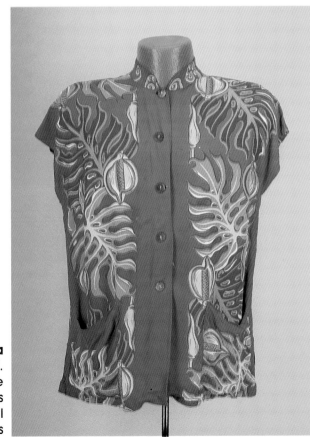

Label: (green) **Kamehameha made and styled in Hawaii**. Lady's "tea-timer" shirt in blue with great large leaves design, Rayon, coconut shell buttons and two waist pockets

Label: **Waikiki Kasuals made in Hawaii**. "Holuku" dress of black cloth with colorful Hawaiian leaves design, Rayon. These dresses with trains were sometimes used as wedding or formal gowns.

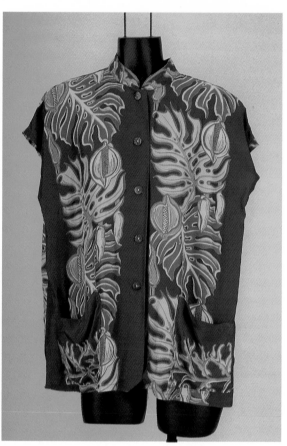

Label: **Shaheen's of Honolulu Made in Hawaii**. "Pake muu" dress of vertical floral design on black background, Rayon

Label: **Kamehameha**. Lady's "tea-timer" in purple with brilliant floral design, Rayon, coconut shell buttons and two watst patch pockets

Label: **Kamehameha**. Dress in purple with brilliant floral design, Rayon

Label: **Kamehameha**. Three-piece swimming oufit of yellow
background with black and grey Hawaiian floral design, cotton

44

Label: **Sundek**. Blue background with colorful vertical floral design, cotton, brown plastic buttons

Label: **Ui-Maikai RN 26300 Made in Hawaii**. White with red vertical floral design, cotton bark cloth, white plastic buttons

Label: **Royal Hawaiian Made and styled in Hawaii**. Dark blue with white floral design in vertical stripes, cotton, coconut husk buttons

Label: **Registered Creative Edge Trademark RN 91855 Made in Hong Kong.** Black with vertical stripes of floral design, Rayon, brown plastic buttons

Label: **Marlboro Permanent Press Made in Hong Kong.** In original cellophane wrapping, green plaid with stripes of white flowers, cotton, white plastic buttons. Distributed by Marlboro Enterprises Inc. 350 Fifth Ave. New York, NY 10001

45

Label: **Malihini Made in Hawaii**. Shaded brown border print with white floral design, cotton, molded plastic Asian writing buttons

Label: **Richard Douglas, Honolulu, Made in Hawaii**. White with green floral border print, cotton bark cloth, molded plastic Asian writing buttons

46

Label: **Royal Palm Hawaii**. Blue border print with colorful horizontal bands and flowers, polyester, blue plastic buttons

Label: **Kennington Ltd California RN 22199.**
Blue with border print of palm trees and
grass hut, polyester, coconut shell buttons

Label: **Hilo Hattie Made in Hawaii,**
with original hang tag **Hilo Hattie...,**
(808) 524-3966. Black background
with colorful bird of paradise floral
design, polyester, black plastic
buttons

Label: **Ul-Maikai RN 26300, 100% cotton, Made in Hawaii.** Dark blue with red floral design, cotton, white plastic buttons

48

Label: **Made in Hawaii.** Selvedge: **MIA Designs-Honolulu Permanent Press.** Shaded red background with large white flowers, cotton bark cloth, molded plastic Asian writing buttons

Label: **Pomaré Hawaii.** Colorful all-over floral design, polyester, white plastic buttons

Label: **Maluna Hawaii**. Dark blue with colorful tropical flower design, polyester, white plastic buttons

Label: **Island image. Made in China.** Colorful floral and mountain scene, Rayon, blue plastic buttons

49

Label: **Paradise Hawaii Made in Honolulu.** White and turquoise large floral design, cotton, silver molded plastic buttons

Label: **Made in Hawaii,** Selvedge printed: **G.V.H. Hawaiiprint.** Blue with green floral design and lettering Hawaii, cotton bark cloth, molded Asian writing plastic buttons

Label: **Paradise Found Hawaii.** Purple with colorful parrots and leaves design, cotton, wooden buttons

Label: **Hoaloha Made and Styled in Hawaii.** Woman's Tea Timer shirt with Mandarin collar, cuffed short sleeves, two patch pockets. Dark blue with light blue flower pots and white orchids design, cotton, coconut husk buttons

Label: **Kula Bay Tropical Clothing Co. Honolulu, Hawaii**. Light blue background with pink parrots and hibiscus flowers, cotton, coconut shell buttons

Label: **Ui-Maikai RN 26300 Made in Hawaii**. Blue with colorful floral design, cotton, brown plastic buttons

Label: **Ui-Maikai RN 26300 100% All cotton, Made in Hawaii**. Selvedge: © **Polynesian Textile**. Woman's long dress and shirt set. Purple background with pink and white floral design, molded plastic buttons

51

Label: **Go Barefoot in Paradise Inc. Honolulu, Hawaii.** Shaded brown background with white floral design, cotton, brown plastic buttons

Label: **Mauna Kea 100% Nylon Made in Korea RN 49028.** Shaded yellow and brown background with palm and floral design, brown plastic buttons

52

Label: **Royal Hawaiian Made and Styled in Hawaii.** Swim trunks. Brown background with white floral design, cotton

Label: none. Black background with colorful orchid and bird of paradise floral design, cotton, white plastic buttons

Label: **Made in Japan for Orchids of Hawaii.** Purple with large white floral design, Rayon, white plastic buttons

53

Label: none. Black with pink and green floral design, cotton, white plastic buttons

54

Label: **RN 72260**. Black background with colorful floral design, cotton, white plastic buttons

Label: **Paradise of the Pacific by Frank**. Grey clouds background with palm trees and boats design, Rayon, brown plastic buttons

Label: **Towncraft Par Excellence Penneys**. Blue with white flowers and palm tree trunks (?), cotton, molded plastic buttons

55

Label: **Made in Hawaii**; Selvedge: **TransPacific Textiles Ltd. D/# DEX-691**. Turquoise blue background with orchid design, Rayon, white plastic buttons

Label: none. Black floral design on shaded brown background, shoulder epaulets and waist band, two patch pockets, cotton, white plastic buttons

Label: **Made in Hawaii.** Backward printed cloth sewn with the "wrong" side out, Dark brown with "Hawaiian Tropic" floral print, shoulder epaulets, back vent, and two pleated and flapped pockets, wooden buttons

Label: **kona*kai Hawaiian Casuals by Jantzen.** Swim trunks and shirt set. Brown leaf design, cotton, brown plastic buttons

Label: **Pomare Hawaii**. Brown and orange floral design, polyester, white plastic buttons

Label: **Pomare/Tahiti Honolulu, Hawaii RN 37145**. Orange with white and yellow floral design, polyester, white plastic buttons

Label: **Kelii's of Hawaii**. Shaded blue background with a rope of green and white flower blossoms, cotton, molded silver plastic buttons

Label: **Malihini Hawaii**. Chartreuse, black and white oversized floral design, heavy textured cotton, white plastic buttons

Label: **Pacific Isle Creation, Made in Hawaii U.S.A.** Colorful large floral design, slit pocket, polyester, fabric covered buttons

58

Label: **Sears Hawaii.** Green, orange and yellow floral design, slit pocket, cotton, white plastic buttons

Label: **Sears Hawaii.** Blue with shaded yellow and blue floral design, cotton, white plastic buttons

59

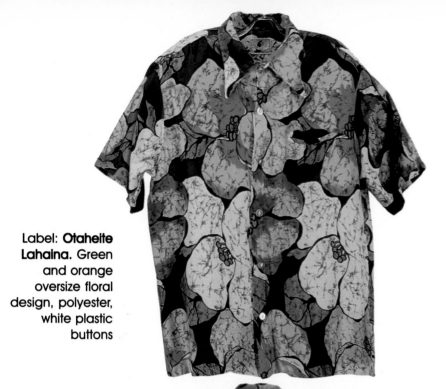

Label: **Otaheite Lahaina.** Green and orange oversize floral design, polyester, white plastic buttons

Label: **Fashioned by Hukilau Fashions Honolulu RN 43220.** Selvedge: **G.V.H. Hawaiiprint.** Green and orange floral design, cotton, white plastic buttons

60

Label: **Keoki's Hawaiian Sportswear Made in Hawaii.** Orange large floral design, cotton bark cloth, white plastic buttons

Label: **Made in Hawaii RN 20942.** Oversize floral design in bold blue, green, brown and white colors, cotton, molded plastic Asian writing buttons

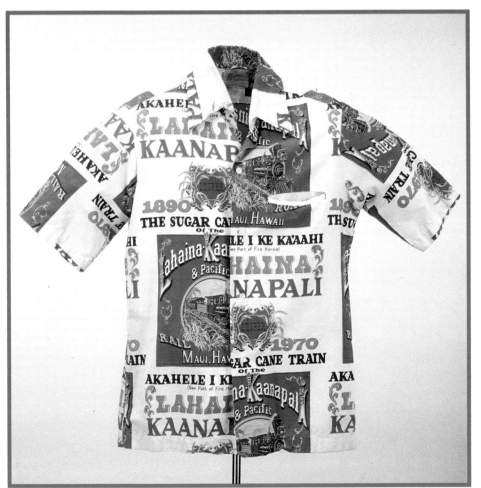

Label: **Keone Sportwear RN 23133 Hawaii**. Colorful "Lahaina-Kaanapali & Pacific Railroad, Maui, Hawaii" 1890-1970 design in rectangles on white, cotton, white plastic buttons

HAWAIIAN SCENICS

Label: **Made in Hawaii RN 37359**. Brown and white rectangles with colorful Hawaiian scenes labeled "Aloha" and "Hawaii," cotton bark cloth, molded plastic basket weave buttons

62

Label: **Made in Hawaii RN 37359**. Blue and white rectangles labeled "Aloha" and "Hawaii" with colorful Hawaiian figures and floral borders, cotton bark cloth, silver molded plastic basket weave buttons

Label: **Fashions of Hawaii**. Green Hawaiian scenes alternating with a vertical band of flower blossoms, cotton, molded plastic Asian writing buttons

Label: none. Shaded blue labeled scenes of places in Hawaii, cotton, coconut shell buttons

Label: **Fashions of Hawaii Made in Hawaii.** Red with Hawaiian scenes in vertical bands alternating with floral design, cotton, molded plastic basket weave buttons

63

Label: **Kai Nani Hawaii**. Turquoise sea with brown thatched huts design, polyester, white plastic buttons

Label: **Napili Made in Hawaii Noniron Fabric**. Volcano design in green and brown tones, polyester, brown plastic buttons

64

Label: **Zuma Beach by Jantzen**. Black with Hawaiian scenes labeled Aloha Waiilea Maui, cotton, brown plastic buttons

Label: **Registered Creative Edge Trademark, Made in Hong Kong RN 91855**. Blue background with colorful Hawaiian images, Rayon, brown plastic buttons

Label: **Pacific Isle Creations Made in Hawaii U.S.A.**
Orange and green landscape with rainbow labeled
Aloha and Hawaii, cotton, fabric covered buttons

Label: **Likeke fashions Hawaii.** Colorful Polynesian
landscape with figures, cotton, molded plastic buttons

66

Label: none. Beige with labeled Hawaiian luau design, cotton, molded chief's head buttons

Label: **Banana Republic Safari & Travel Clothing Co., Made in Thailand RN 54023**. Yellow with Hawaiian scenes, each labeled, Rayon, yellow plastic buttons

Label: **Created in Hawaii for Diamond Head sportswear International Marketplace Waikiki.** Swim trunks. Colorful polynesian design, cotton

Label: **Island Image Made in China**. Blue background with tropical flowers and Hawaiian landscape design, Rayon, blue plastic buttons

Label: **Island Image, Made in Bangladesh RN 39050**. Colorful floral design with Hawaiian landscape, Rayon, blue plastic buttons

68

Label: **Canoes Sportswear Honolulu.** Pastel islands design, Rayon, beige plastic buttons

Label: **Uniforms by Malia.** Light blue background with dark blue HEI electrical company design, 55% polyester, 45% cotton, white plastic buttons

Label: **Active Culture Mix,** 100% cotton, Made in U.S.A. Wash this separately in cold water Tumble dry low. No Bleach Ironing Prohibited By Law. Polynesian landscape print with orange and blue, cotton, pink plastic buttons

69

Label: **Big Kanaka by Corsini.** Island landscape and flowers with shaded brown background, cotton, brown wooden buttons

Above:
Label: (green) **Kamehameha Made and Styled in Hawaii.** Selvedge printed **"Original Matson Menu Design by Eugene Savage."**
Woman's tea-timer shirt with beautiful polychrome "Island Feast"
design (c. 1950) after the mural by Eugene Savage, Rayon,
coconut husk buttons.

Right:
Label: **Kamehameha.** Selvedge printed **"Menu Design by Eugene
Savage."** Lady's "pake muu" dress with Eugene Savage's colorful
"Island Feast" print design, Rayon

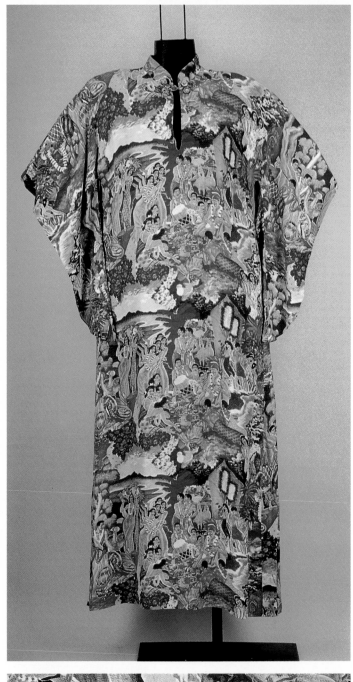

Above:
Label: **Original by Hale Hawaii Made in Hawaii.** Lady's shirt of green and multi-colored "Land of Aloha" print design, Rayon crepe

Left:
Label: **Lauhala Made in Hawaii.** "Pake muu" dress with open-at-the-top sleeves and side zipper, "History of the Islands" print on black background, Rayon

Label: **Original by Hale Hawaii Made in Hawaii.** Dress and short jacket in floral design fabric with waterfall and rainbow

Label: **Made in Hawaii.** Woman's dress. Green background with Aloha Hawaii lettering and polynesian scenes in floral wreaths with gold paint details, Rayon

Label: **Sun Country.** Blue with colorful Island scene, Rayon, plastic buttons

73

Label: **Off Shore, Designed and produced by Daystar of California.** Beige background with colorful Christmas in Hawaii theme design with lettering, cotton, coconut shell buttons

HAWAIIAN SAILING

Label: **Made in Hawaii**.
Selvedge: **GVH. Hawaii
Print**. Colorful palm trees
and sail boat design,
polyester, molded
plastic buttons

Label: **Kal Nanl Hawall**. White and green fragmented tapa and scenic design, cotton, white plastic buttons

Label: **HRH His Royal Highness Hawall**. Placket closure with three buttons. White with brown print of the boats and chart in the Honolulu Race 1971, cotton, wooden buttons

Label: **Royal Hawaiian Made and styled In Hawall**. Swim trunks. Yellow and brown sailboat and floral design, cotton bark cloth

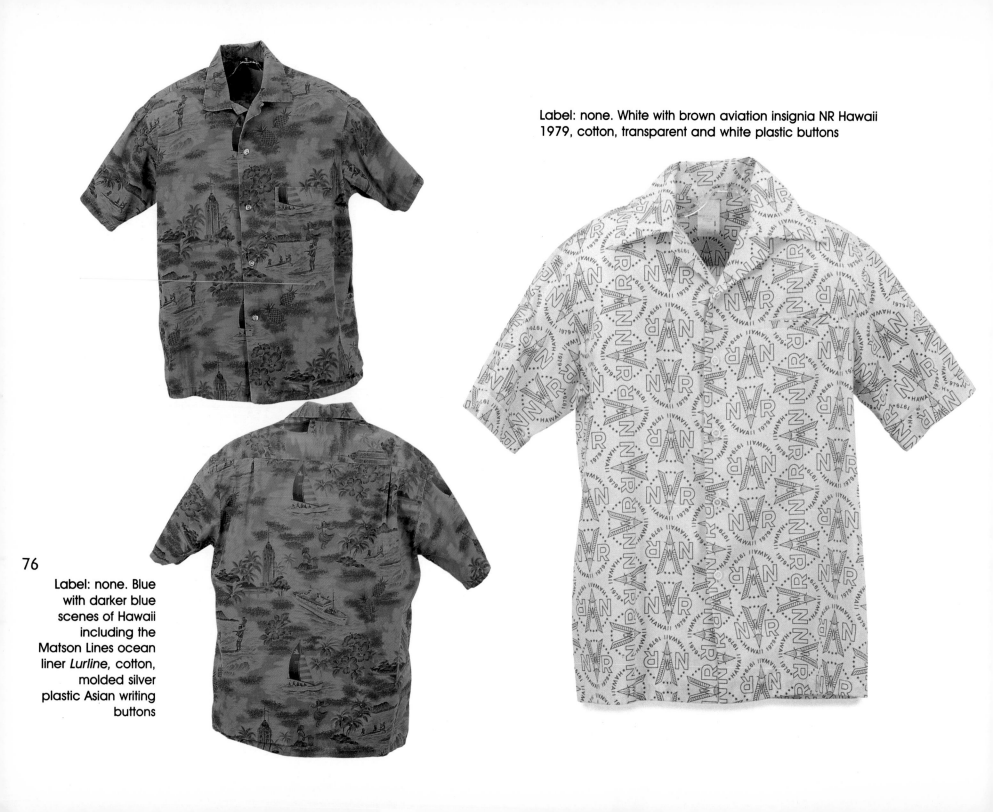

Label: none. White with brown aviation insignia NR Hawaii 1979, cotton, transparent and white plastic buttons

76

Label: none. Blue with darker blue scenes of Hawaii including the Matson Lines ocean liner *Lurline,* cotton, molded silver plastic Asian writing buttons

Label: **Kahala by HRH Hawaii.**
White with blue fish design,
cotton, white shell buttons

Label: **Z*Force Made in Swaziland RN
69347 55% cotton, 45% polyester.** Blue
with fish design, blue plastic buttons

HAWAIIAN FISH

Label: **Lel-O'Hawaii Sportswear**. Blue with fish design and gold paint accents, cotton, molded plastic Asian writing buttons

Label: **Canoes Sportswear Honolulu**. Fish design on grey background, Rayon, tan plastic buttons

Label: **HRH His Royal Highness Hawaii**. Swim trunks. Yellow and brown fish and floral design, cotton

HAWAIIAN FRUIT

Label: none. Colorful pineapple
and plantation border print,
cotton, white plastic buttons

Label: **Malihini Hawaii. RN 22654**. Brown and white border print with pineapple design, polyester, white plastic buttons

Label: **Made in Hawaii.** Blue with darker blue pineapple design and gold painted details, cotton, molded plastic Asian writing buttons

80

Label: **Made in Hawaii.** Child's shirt. Dark yellow shaded background with darker pineapple and fragmented tapa design, cotton, molded plastic Asian writing buttons

Label: **Fashioned by Hukilau Fashions Honolulu RN43220**. red background with pineapples and beige leaves. Rayon. White plastic buttons

Label: **Ocean Pacific sunwear**. Long sleeves. Sewn with the "wrong" side out. Beige background with colored pineapples, polyester, coconut buttons

Label: **Angels Flight, Made in Bangladesh**. Blue background with floral, fish, guitar, and surfer design, Rayon, brown transparent plastic buttons

Label: None. Dress with altered and shortened sleeves of fabric with green background and bananas design, Rayon

82

HAWAIIAN BEER & COCKTAILS

Label: none. Brown background with Primo Hawaiian beer insignia and map of Hawaii design, cotton, molded plastic Aloha Hawaii buttons

Label: **Ui-Maikai Made in Hawaii**; yellow with green lettering. Pea green Primo Hawaiian Beer design, cotton bark cloth, molded plastic buttons

Label: **Duke Kahanamoku Made in Hawaii by Kahala.** White with labeled rows of cocktail glasses, cotton, molded crown design metal buttons

Label: **Honolulu Gold.** Blue landscape and cocktail glasses with recipes, cotton, coconut shell buttons

84

CALIFORNIA

Label: **Sun fashions Sportswear styled in California.**
Red with dancers and
pineapple stylized design,
cotton, white shell buttons

Below:
Label: **Berkley shirtmakers RN 18798 Made in U.S.A.** Blue with orange and white Exeter California Oranges design, cotton, white shell buttons

Right:
Label: **Berkley shirtmakers RN 18798 Made in U. S. A.** White with orange and blue Exeter California Oranges design, cotton, white shell buttons

86

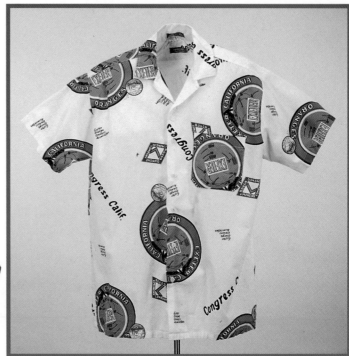

Label: I.N.C. International Concepts worn worldwide for comfort and style, Single needle, 100% Rayon, Made in Korea WPL 8046. Blue with labeled scenes of Los Angeles, California, white shell buttons

Label: **Ocean Pacific**. Colorful abstracted grass huts design labeled Pacific Ocean with rubber stamp markings, cotton, white plastic buttons

Left:
Label: **ocean pacific RN 52319**. Beige with palm trees and rubber stamp shipping labels, and lettering Pacific. Cotton, beige plastic buttons and the original paper hang tag

Below:
Label: **California Rainbow**. Shaded pink background with blue floral design, Rayon, white plastic buttons

88

Label: **Barefoot in Paradise RN 36979 Made in U.S.A. ©** **Disney.** Mickey Mouse characters with palm trees, cotton, brown plastic buttons

Label: **Mickey & Co. by J. G. Hook Made in Guatemala ©** **Walt Disney Co.** White knit collar and light blue background with colorful Mickey Mouse with various beach toys, cotton, grey plastic buttons impressed "Mickey & Co."

90

Label: **Papaya product shirts Made in U.S.A.**
Pastel Tarzan fighting scenes in rectangular
frames, Rayon, white plastic buttons

Label: **Michael Seroy Made in U.S.A.** Swim trunks and shirt set. Colorful Tarzan wrestling animals print, Rayon, white plastic buttons

Label: **Created in the beach for Big Waves by Heet Sportswear Always be casual.** Yellow "Harrison Ford The Mosquito Coast" lettering and island border design, curved shirt tail, cotton, yellow plastic buttons

91

Label: **B.S.R. Made in Turkey RN 64371**. White with fruit menu design, heavy cotton, white plastic buttons

U.S.A. MAINLAND

Label: **B.S.R. Made in Costa Rica RN 64371**. White with colorful fruit menu design, cotton, white plastic buttons

Label: **All Cotton, Single Needle Tailoring Made In U.S.A.** Turquoise with colorful fruit design, turquoise plastic buttons

Label: **Original Pointzero concept.** Square border print with white center and colorful fruit in the border, cotton, clear plastic buttons with black stripes

93

Label: **Van Cort sport shirt Made in Burma RN 14711.** Neutral background with orange and black coconut liquor "Tropical taste" design, Rayon, white plastic buttons

Label: **Corona Extra Clothes presented by Daniel Hilu officially licensed by Pamapromex Made in Mexico.** White with yellow and blue Corona beer bottles and trademark design, cotton, white plastic buttons

Above:
Label: **Sergio Valente, Made in United Arab Emirates.** Black background with beer bottles design, cotton, brown plastic buttons

Left:
Label: **Sara Toga Sport RN 81027 Zack Apparel Company Made in U.S.A.** Red with Campbell Soup advertising characters in tropical setting, cotton, red plastic buttons

Left:
Label: **marc daniels Made in U.S.A.** Colorful fast food drive-in design, cotton, white shell buttons

Below:
Label: **Paragraff, Made in India RN 75035.** Colorful labeled scenes of American popular life and coins dated 1990, Rayon, white plastic buttons

95

Label: **The Izod Club RN 21008 Made in Hong Kong**. White with colorful golfers, cotton, white shell buttons

Label: **Mardac ® poplin Permanent Press 65% Dacron ® polyester, 35% combed cotton**. American Legion seals and colorful floral design, white plastic buttons

Label: **Big Dogs Santa Barbara Made in U.S.A.** Blue with palm tree and monkey design with black and white dogs, cotton, black and white speckled buttons

Label: **Heet Sportswear.** White with colorful musicians and dancers, cotton, white plastic buttons

Label: **Bullock and Jones San Francisco.** Red background with Jazz lettering, music and floral design, cotton, white shell buttons

97

Label: **Orvis Made in U.S.A.** White with black newspaper advertisements for fishing supplies, cotton, grey metallic plastic buttons

Label: **Sportswear, 100% Rayon, Hand Printed, Imported by S.S.Kresge Co., Made in Japan.** Colorful fishing scene, white plastic buttons

98

Above:
Label: **Since 830 Woolrich Rugged Outdoorwear, 100% cotton Assembled in Mexico of USA components**. White with green fishing equipment design, heavy cotton, white plastic buttons

Right:
Label: **BAIK**. Black with green fish and colorful small floral design, Rayon, black plastic buttons

Above:
Label: **Woolrich since 1930 Rugged Outdoorwear**. Army green background with fishing equipment in tan, cotton, white shell buttons

Left:
Label: **Since 1830 Woolrich Rugged Outdoorwear Made in U.S.A.** Grey background with fishing equipment design, cotton, white plastic buttons

Label: **Topflight.** Shaded
green background with red
lobsters design, cotton,
green plastic buttons

Label: **Verte Vallee Just a
natural product Made in
Hong Kong RN 80791.**
Long sleeves, blue
background and colorful
scenes of the Hanging
Rock Battlefield, Southern
Railroad and the town of
Kershaw, South Carolina,
Rayon, one glass Verte
Vallee button and the
remaining ribbed white
plastic buttons

FLORIDA

Label: **Hutspah Made in U.S.A.**
White with colorful map of
Florida and scattered beach
scenes, cotton, white plastic
buttons

102

Label: none. Blue background
with imaginary space ships and
astronauts design, Rayon, brown
coconut shell buttons

Label: **Kennington Ltd. California**.
Red astronauts design, Rayon,
coconut shell buttons

Label: **Kennington Ltd
California Made in China**.
Blue astronauts design,
Rayon, coconut shell buttons

Label: **Substudio.** colorful horizontal beach crowd design, cotton, yellow plastic buttons

Label: **Le Tigre' Made in U.S.A.** Island scene with sailboat, girl on beach and hut in design, cotton, blue plastic buttons

103

Label: **King Arthur vacation wear 100% polyester RN 40832 Made in Korea.** Shaded brown background and waterfall with colorful rainbow and floral design, polyester, white plastic buttons

Label: **Tommy Hilfiger Made in Thailand, Fabrique en Thailande, Hecho en Thailandia.** Blue with colorful floral design, Rayon, white shell buttons

104

Label: **Palm Tree of Branford RN 44921 Made in Korea.** Green crackle background with green and brown floral design, polyester, white plastic buttons

Label:
**J.J.Cochran
Made in U.S.A.**
Pink back-
ground with
blue tropical
village scene,
brown plastic
buttons

Label: **Tommy
Bahama quality
since 1993 Made
in Hong Kong**, and
separately: **Tommy
Bahama Relax.** Tan
with blue and
yellow large floral
design, cotton,
coconut shell
buttons

Label: **Tommy
Bahama Quality
since 1993 Made in
Hong Kong.** Shaded
grey background
with floral design,
Rayon, coconut shell
buttons

105

Label: **Tommy Bahama
Quality since 1993,
Made in Hong Kong.**
Island landscape and
bird of paradise flower
design on green-grey
background, 60%
Rayon, 40% cotton,
coconut husk buttons

Label: **Campus knit shirt 100% nylon Made in Korea**. Grey background with colorful jockeys on horseback design, white plastic buttons

Label: **Club Room Made in Malaysia WPL 8046**. Knit polo shirt with placket neck closure, grey background with colorful floral design, cotton, two brown plastic buttons

106

Label: none. Button down collar, blue with pink and light blue floral design, cotton, coconut shell buttons

BERMUDA

Label: **English Sports Shop Bermuda**. Light blue with map of Bermuda and colorful scenes, Rayon, white plastic buttons

Label: **Capital Bermuda**. White with dark blue scene in Bermuda, 65% polyester 35% cotton, white plastic buttons

Label: **English Sports Shop, Bermuda**. White with liquor bottles in a horizontal line, cotton, white plastic buttons

Label: **Orvis Made In U.S.A.** White with a chart of the ocean near Bermuda Island, cotton, wooden buttons

108

Label: **Capital Bermuda.** White with blue map of Bermuda and sites of the island, 65% polyester 35% cotton, white plastic buttons

BAHAMAS

Label: **Hand Printed Original by Bahamas Mfg. Co.** and separately, **Made in the Bahamas.** Colorful Bahamas Islands map, floral and fish design, cotton, white plastic buttons

Label: **The Islander, Bahamas.** Blue with liquor bottles design, 65 % polyester, 35% cotton, blue plastic buttons

Label: **Maas Brothers of Florida.** Beige with colorful liquor bottles in horizontal rows, buttoned waist tabs, cotton, white plastic buttons

Label: **Windswept in the Bahamas.** White with colorful map of the Bahama Islands and fish design, cotton, white plastic buttons

CARIBBEAN

Label: **Kudos**. White with colorful
maps of the Caribbean islands,
cotton, white plastic buttons

Left:
Label: **-Lo Shop Antigua, West Indies,** and separately **Certified by West Indian Sea Island Association (Inc.) cotton.** Flags of the West Indian nation islands: St. Vincent, Guyana, Jamaica, Monserat, Puerto Rico, ..., cotton, white plastic buttons

Below:
Label: none. White with colorful Jamaican places and activities, cotton, brown plastic buttons

Label: none. Swim trunks. Blue background with flags of the Caribbean nations, cotton

Right:
Label: **Modesto Garments of Distinction Kingston, Jamaica**. Border print of Jamaican people in front of grass huts, cotton, white plastic buttons

Below:
Label: **Sportswear by Patrick Washable**. Jamaica map and activities on light blue background, cotton, white plastic buttons

113

Label: **Sun Island.** Shaded grey background with sailfish and red floral design and lettering "Jamaica," cotton, white plastic buttons

114

Label: **Modesto Garments of Distinction, Kingston, Jamaica.** Colorful floral design with labeled Jamaican places, Rayon, white plastic buttons

VIRGIN ISLANDS

Label: **Java wraps St. Croix, St. Thomas, St. Maarten,** and separately, **Made in Indonesia**. Black with bamboo design, Rayon, coconut shell buttons

Below left:
Label: **St. Thomas Stationers U.S. Virgin Islands.** White with vertical rows of blue labeled buildings in St. John and St. Thomas, cotton, white plastic buttons

Below:
Label: **nautica Made in Macau RN 67835.** Map and activities of the Virgin islands, cotton, white shell buttons

Label: none. Blue background with colorful map of Guadeloupe Island and its famous products, cotton, blue plastic buttons

WEST INDIES

Right:
Label: **The Bagshaws of St. Lucia W.I.** Zippered front closure, white with green sailing ships border design, two slit pockets, cotton

Below:
Label: none. Red with pineapple border print and dancing woman labeled "Martinique, Rocher du Diamant," cotton, white plastic buttons

117

118

Label: **The Bagshaws of St. Lucia W.I.** Zippered front closure, beige with colorful geometric design only on the front, two slit pockets

Label: **Bajan Made in Barbados**. Open neck style with white
binding and two waist patch pockets. Orange with island
scenes labeled "Barbados Island of Love", polyester, no buttons

Label: **Sun Bay**. White with blue blueprint drawing of a ship, "Champion Racer Plans," cotton, white plastic button

Label: **Bajan.** Colorful "Caribbean Paradise" design with brown and white rectangles and floral borders, ribbed cotton, white plastic buttons

Label: **liberty Garments Made in Grenada, W.I.** Blue with red beach huts design labeled "Grenada Isle of Spice," cotton, white plastic buttons

AFRICA

Above:
Label: **Active Culture Mix...** blue with hieroglyphics in stripes, cotton, blue plastic buttons

Right:
Label: **Fashion Seal Shane**. Brightly colored landscape with women carrying bowls on their heads, cotton, purple plastic buttons

Label: none. White background with green and black design including a photo-reproduction of Houphouet Boigny, President and founder of the RDA-PDC I political party, 1905, cotton, black plastic buttons

122

Label: **Z*Force Made in Swaziland RN 69347, 55% cotton, 45% polyester**. Red with map of the Seychelles Islands and floral design, red plastic buttons

CLOTHING LABELS

Manufacturers and retailers are identified by their labels on the clothes.

Capital
BERMUDA

Cooke Street®
Honolulu
R/NO 20957
MADE IN U.S.A.
55% POLY 45% COT
STYLE# 2002LZ

ENGLISH SPORTS SHOP
BERMUDA

HAWAIIAN HOLIDAY
SPORTS WEAR
MADE IN HAWAII

HRH
HIS ROYAL HIGHNESS
HAWAII

Carlton Hill XL
PERMANENT PRESS
100% POLYESTER
MACHINE WASH TUMBLE DRY
RN 39012   MADE IN KOREA

Corona
Extra
CLOTHES
PRESENTED by
DANIEL HILU
OFFICIALLY LICENSED BY
PAMAPROMEX
MADE IN MEXICO

Fashion Seal
Shane
-L-
MACHINE WASH WARM-TUMBLE DRY
100% COT SANF
70366 55040

HET
SPORTSWEAR
L
100% COTTON/
COTON
MADE IN/FAIT EN
HONG KONG
RN 60907

Fashioned by
Hukilau Fashions
HONOLULU RN43220

Champion
Westerns
PERMANENT PRESS
17 - 35

100% RAYON
RN 91855
MADE IN HONG KONG
L
SEE REVERSE FOR CARE
Registered
CREATIVE EDGE®
Trademark

Fashions
of Hawaii
MADE IN HAWAII

Helena's
MADE IN HAWAII, U.S.A.
65% POLYESTER
35% COTTON
STYLE 192
8182

Fashioned by
Hukilau Fashions
HONOLULU RN43220

XL
COTTON
MADE IN MEXICO
CLUB
R
ROOM

Dael's
STYLE1049
L

Fashions
of Hawaii
MADE IN HAWAII

Hilo
Hattie
MADE IN HAWAII
RN 37145
LOT 905
STYLE 560
100% POLYESTER

by Fashioned
HUKILAU FASHION
HONOLULU RN43220
Machine wash
Cold water, mild
soap. Tumble dry.
Do not bleach,
soak, or wisst.

100% COTTON
MADE IN U.S.A.
J. J.
COCHRAN
MADE IN U.S.A.
XL

DIVI-DIVI
CURSOU

GO
BAREFOOT
In PARADISE
Inc. HONOLULU, HAWAII

MADE IN ITALY   GB 40
F3
HOM
Création France
D5   USA 40
FABRIQUÉ
EN ITALIE

Hutspah
MADE IN USA

CoCo Shop
antigua w.i.
west indian
sea island cotton
L

Duke Kahanamoku
MADE IN HAWAII BY KAHALA
S

ORIGINAL
by
Hale Hawaii
14   MADE IN
HAWAII

Honolulu
Gold

Jantzen
MADE IN U.S.A.
42
SEE REVERSE FOR CARE

124

125

Machine wash,
cold water
mild soap,
Tumble dry
Do not bleach,
soak or twist.
MADE IN HAWAII
RN 37359

Maluna
Hawaii

MICHAEL SEROY
MADE IN U.S.A.
S

nautica L/G
MADE IN MACAU
FABRIQUE A MACAO
100% COTTON POPLIN
100% COTON POPELINE
CA 07615

OP
ocean pacific
sunwear
XL
CARE INSTRUCTIONS
ON REVERSE SIDE

UNIFORMS BY **Malia**

marc daniels
17-XL-17½
MADE IN U.S.A.

Mickey & Co.
by J.G. HOOK
XL
100% COTTON
MADE IN
GUATEMALA
CARE ON REVERSE

NEW
HORIZON
MADE IN HAWAII
of Imported Fabric.
Machine wash cold.
Mild soap. No bleach.
Tumble dry cool. Use
cool iron only.

OP
ocean pacific

MALIHINI
HAWAII

65% DACRON® POLYESTER
35% COMBED COTTON
MARDAC
POPLIN
PERMANENT PRESS
M
"OVER"

Active Culture.
100% **MiX** MADE IN U.S.A.
COTTON
WASH THIS SEPARATELY IN COLD WATER
TUMBLE DRY LOW--NO BLEACH
IRONING PROHIBITED BY LAW

NEWPORT B·L·U·E™

Orchid Fashions
MADE IN HAWAII
65% Polyester
35% Cotton
Size Style
L

MALIHINI
HAWAII

MAUNA KEA
XL

MODESTO
Garments
OF DISTINCTION 6
KINGSTON JAMAICA

Nui Nalu
HAWAII
MADE IN USA
SIZE 12 STYLE 203
Lihue

MADE IN JAPAN FOR
Orchids of Hawaii

126

M MALIHINI HAWAII
FOR
LIBERTY HOUSE
RN 28004
100% POLYESTER

Merona™
ALL COTTON
MADE IN
SINGAPORE
RN 39719
XL
CARE ON REV.

Mukai Fashions
MADE IN HAWAII

OFF SHORE®

ORVIS®
MADE IN U.S.A.
'est fish

M MALIHINI HAWAII
DESIGNER'S COLLECTION
liberty house
OF HAWAII

Michael Geraldino
HAND SCREENED
Tailored in China

NO IRON FABRIC
NAPILI
Made in Hawaii

ON THE BRINK

ORVIS
L
100% COTTON
MADE IN U.S.A.
on information from
1:60,500 Ed. 1949
950
1000

127

ALOHA!